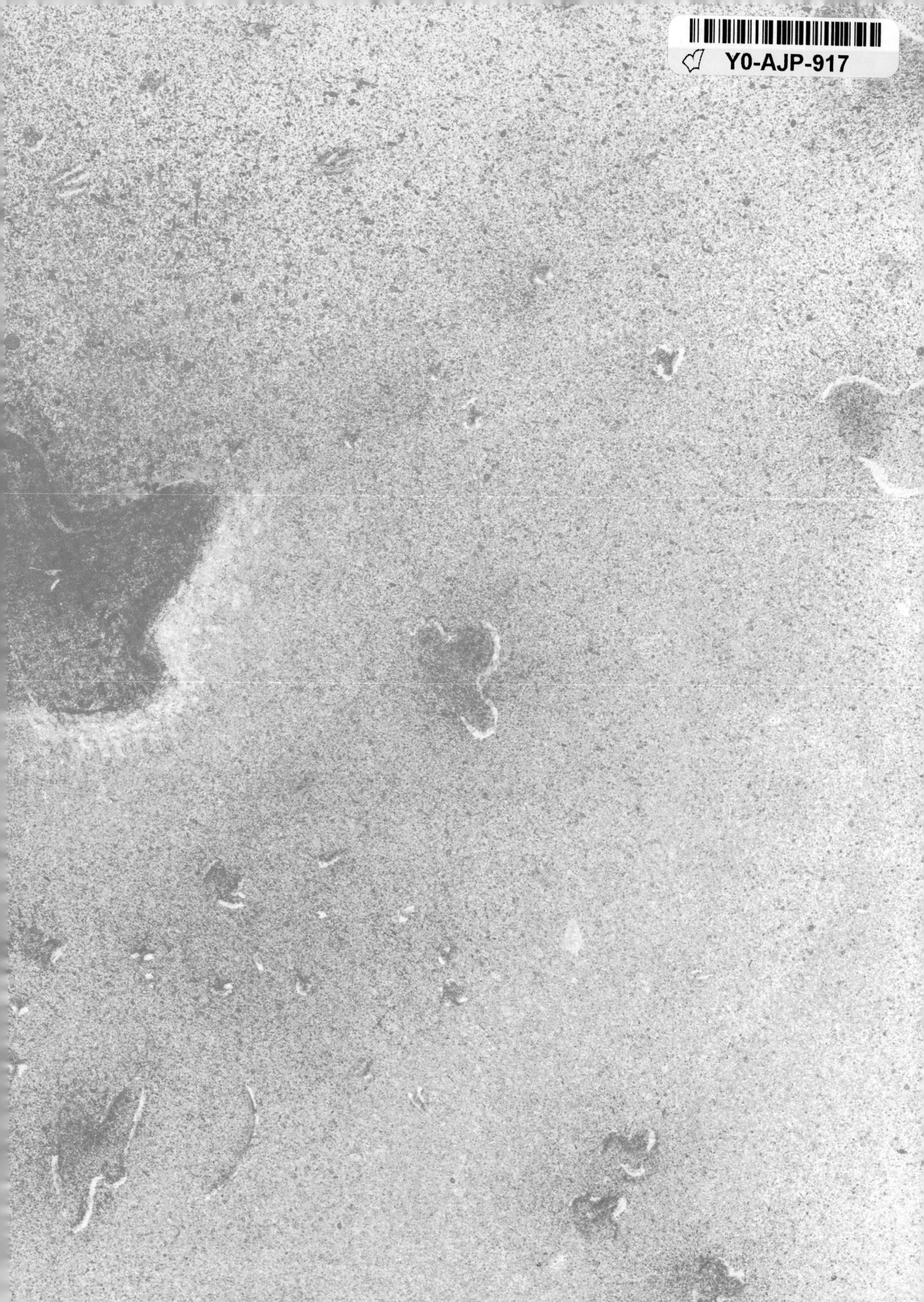

MUTTABURRASAURUS
AN AUSTRALIAN DINOSAUR IN ITS TIME AND SPACE

MARY E. WHITE

Illustrated by
ROBYN MUCHE

Houghton Mifflin Australia

GEOLOGICAL TIME COLUMN

ERA	PERIOD	AGE IN MILLION YEARS		
CAINOZOIC	QUATERNARY	1.6		AGE OF HUMANS
CAINOZOIC	TERTIARY	66.4	FLOWERING PLANTS	AGE OF MAMMALS
MESOZOIC	CRETACEOUS	144		← MUTTABURRASAURUS
MESOZOIC	JURASSIC	208	LIFE ON THE LAND	AGE OF REPTILES
MESOZOIC	TRIASSIC	245		
PALAEOZOIC	PERMIAN	286		
PALAEOZOIC	CARBONIFEROUS	360		AGE OF AMPHIBIANS
PALAEOZOIC	DEVONIAN	408		
PALAEOZOIC	SILURIAN	436		
PALAEOZOIC	ORDOVICIAN	505	LIFE IN THE SEA	AGE OF FISHES
PALAEOZOIC	CAMBRIAN	570		MARINE CREATURES WITH SHELLS

QE 862 .D5 W65 1990

INTRODUCTION

Dinosaurs capture the imagination. For most people, mention of prehistoric animals and fossil bones immediately conjures up images of monsters and the Age of the Reptiles, and of a strange and ancient world very different from our own.

Australia has its dinosaurs, and, in particular, an all-Australian one—the MUTTABURRASAURUS—which lived about 120 million years ago. The world in which it lived was one of rivers, billabongs, swamps and plains, very much like parts of our continent today. But the vegetation was composed of plants different from those of today's scrub and forest. The plants of the time of the Muttaburrasaurus were of ancient groups—the ferns, conifers, cycads, ginkgos, clubmosses and horsetails—all of which have very long fossil histories. Some very early ancestral flowering plants were beginning to spread amongst these ancient varieties, but they were rare.

The land animals alive at the time formed a rich and varied community, including dinosaurs of all sizes, other reptiles, early varieties of birds, and many insects and other invertebrates. Fish, amphibians, turtles and crocodiles lived in the watery places. Reptiles dominated the animal world, and mammals, like flowering plants, were rare.

Mammals and flowering plants dominate the animal and plant kingdoms today. They only became abundant after the dinosaurs had become extinct. This occurred at the end of the Cretaceous Period of geological time, about 65 million years ago, when there were many changes and the modern world was starting to emerge.

We read the story of prehistoric plants and animals by studying fossils preserved in rocks. From the changing animals and plants in the fossil record we can see how evolution has taken place through the ages, from the very beginning of life, through all the later stages. From the fossil evidence, and from our knowledge of the living world of today, we can reconstruct the animals and environments of the distant past.

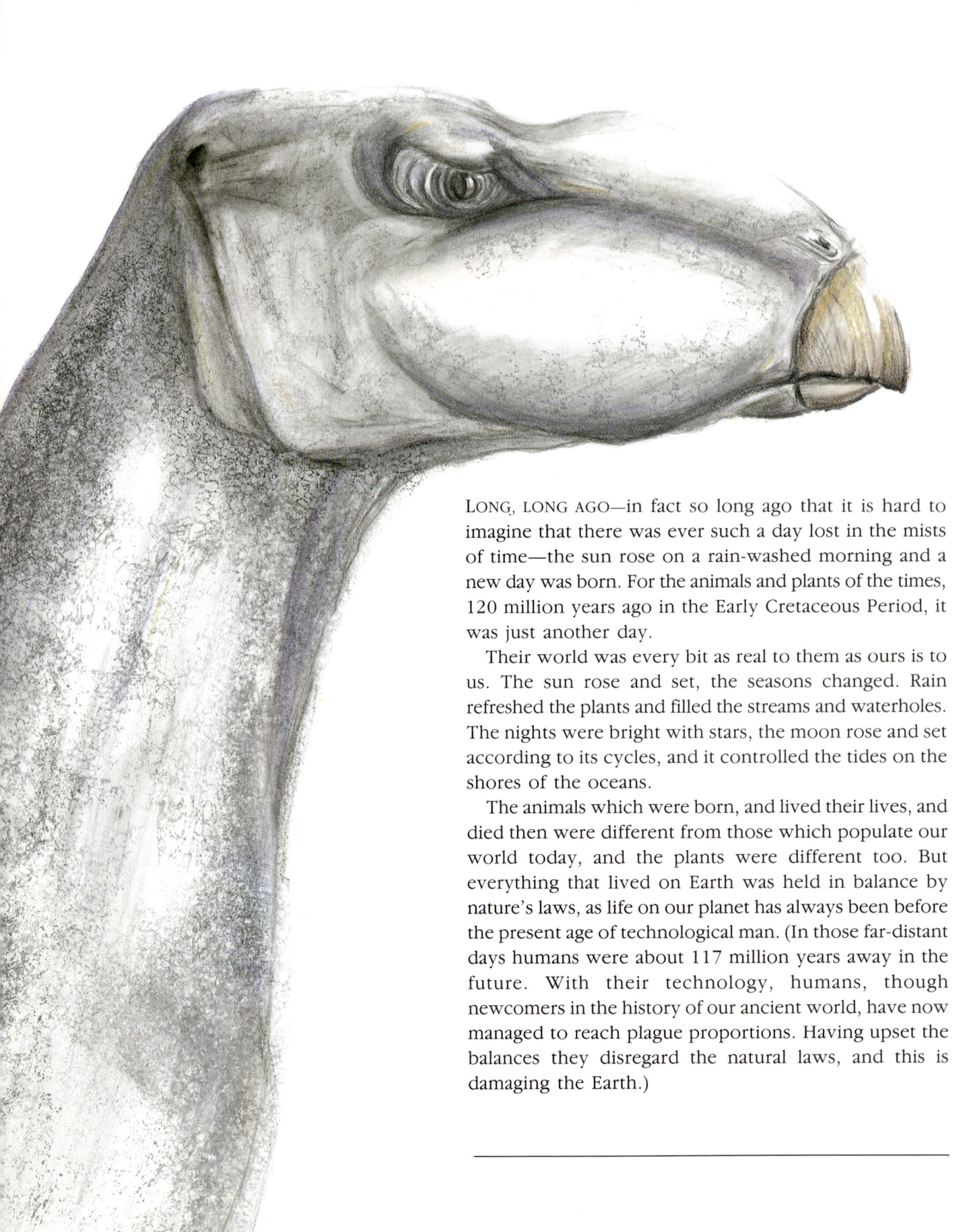

LONG, LONG AGO—in fact so long ago that it is hard to imagine that there was ever such a day lost in the mists of time—the sun rose on a rain-washed morning and a new day was born. For the animals and plants of the times, 120 million years ago in the Early Cretaceous Period, it was just another day.

Their world was every bit as real to them as ours is to us. The sun rose and set, the seasons changed. Rain refreshed the plants and filled the streams and waterholes. The nights were bright with stars, the moon rose and set according to its cycles, and it controlled the tides on the shores of the oceans.

The animals which were born, and lived their lives, and died then were different from those which populate our world today, and the plants were different too. But everything that lived on Earth was held in balance by nature's laws, as life on our planet has always been before the present age of technological man. (In those far-distant days humans were about 117 million years away in the future. With their technology, humans, though newcomers in the history of our ancient world, have now managed to reach plague proportions. Having upset the balances they disregard the natural laws, and this is damaging the Earth.)

Suppose for a moment that we can become time-travellers, able to move back and forth in the past, to any precise time we might choose, and that we can experience that rain-washed morning in the day of the Muttaburrasaurus. We will hover in a time machine above the ancient landscapes and land where we choose on the Earth of long ago. We will look at its inhabitants without being seen.

As we come out of the black void, dawn illuminates the world. Just as the modern world appears to satellites from space, half of the Early Cretaceous globe appears below us. It looks different from the world we know . . . There lies the great Southern Land—GONDWANA—in which South America, Africa, Madagascar, India, Antarctica, Australia and New Zealand are all joined. Much of the land is covered by shallow seas, but we can trace the outlines of the supercontinent against the darker blue of the deep oceans.

Over the next 80 million years Gondwana will split up and the pieces will drift apart. Shallow seas will advance and retreat over different parts of the land, and by the time we are born into this amazing world, the continents will be in their present-day positions on the globe. Even now the continents are still on the move, and Australia is travelling northwards at about seven centimetres a year. It will collide with Asia in about 60 million years and will push up a new mountain range as high as the Himalayas. Our journey back to the Early Cretaceous, to the part of the ancient supercontinent which is later to become Australia takes us towards the South Pole, because Australia was attached to Antarctica at this time.

As we cross the Australian territory we see that the continent is flooded. There is a vast inland sea—the Eromanga Sea—covering about half of the land, and the place to which we are heading is on the edge of this sea, somewhere between Longreach and Hughenden on the modern map of Queensland. Soon we are close enough to see details of the landscape below.

Behind the sand dunes and beaches of the inland sea lie huge swamps. They are lush and green, densely covered in vegetation. As the morning warms, hazy mists rise above the stagnant pools which are green with algae. There are channels where the water is flowing between well-defined banks; a maze of trickles, rivulets and little streams. Swamps like these with dense plant growth existed for many millions of years during the preceding Jurassic Period of geological time as well. The warm and wet climatic conditions resulted in luxuriant plant growth, and the decayed vegetable matter which accumulated in the swamps produced the coal deposits which are mined today in northeastern New South Wales and southeastern Queensland. Up ahead, across the steamy swamplands, is what we have come so far to see. Where a river falls over rocks into a wide pool with sandy edges is a swampy area and a herd—or "burra"—of Muttaburrasauruses. A rocky ledge separates their reach of the river from the plain which lies beyond.

It is easy not to notice how large these creatures really are. When we think about them as "prehistoric" we imagine them as larger-than-life. But here, alive in their own time, they are elegant and fascinating and in tune with their environment.

The burra of nine individuals owns this pleasant territory. The huge male Muttaburrasaurus lying over there in the mud is having his neck nuzzled by his favourite female, and he is completely at ease. He has a body as big as a cart-horse, and his thick tail is as long again. A flexible neck and smooth head bring his length to about seven metres from his lumpy nose to the pointy tip of his tail. He is a sort of purplish black colour, except for his underside, which is a delicate silvery pink. He looks as though he is made of soft vinyl—with a shadow-scaly finish like an expensive snake-skin handbag. He has a marvellous face, with great, sad, hangdog eyes, a swollen nose, and a sort of beak. His favourite female and the three

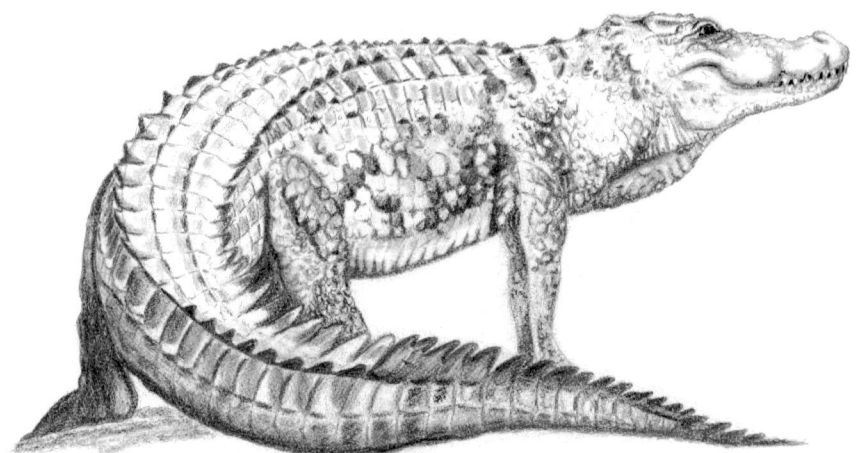

other females in his breeding group are smaller than he is, paler in colour and iridescent. These dinosaurs are "bird-hipped" and they walk with a hip-wiggling, undulating waddle. They use their elegant hands to gesture at each other.

Their conversation sounds like a chorus of didgeridoos. The noises they make echo about in the hollow chambers of their lumpy noses, and resonate and burble and rumble, sounding just like a corroboree. They snap their beaks and make tapping noises to add to the effect.

The four other members of this burra are all juveniles. They are playing in the water, swimming, duck-diving and slapping the water with their tails. Over there, one is climbing out onto a rock to rest. His colouring and markings are different from the adults', and he looks as though his skin is too loose for him. It is all folds and rumples. He has a purple stripe along his back, and a pattern of tabby-cat stripes and spots all over—useful camouflage when he is in the scrub or forest.

As the heat of the day increases, the adults, one by one, enter the water of the deep pool and submerge themselves until only their bulbous noses and eyes are exposed, like hippos in an African river. Dragonflies flit over the water, cicadas are singing in the bushes, and there is obviously not going to be any activity here for a while. But on either side of the wide, pleasant waterway are park-like river flats which are worth exploring.

They are dotted with cycads which look like palms, with tree-ferns, and occasional big trees. Some of the trees are of the maidenhair family, like *Ginkgo* which is their only living relative; most are conifers. Small bushy plants form the understorey and there are bracken ferns everywhere. In wetter places there is a springy mat of mosses, and there are bushes with ferny leaves, small cycads like our burrawangs, and spiky, grass-like horsetails with segmented stems like bamboos.

There are no flowers and no grasses, for the flowering plants are only beginning to evolve at this time and will not be obvious in the vegetation for another ten million years yet. The time of the Muttaburrasaurus is still the age of the conifers, cycads and ferns and not of the flowering plants, just as it is the age of reptiles and not of mammals.

Using our time machine to move backwards and forwards in space and time, we can take a look at anything particularly interesting in this world of long ago. In the bushes below a grisly sight awaits . . .

A flock of small bird-like dinosaurs has been feeding on an overnight kill. They are peculiar creatures, which look like a cross between a featherless chicken and a frill-less frilled lizard. They run on their hind legs, with their small front legs folded back like under-developed wings. They stretch their necks forward while they run. Their pointed heads have snouts rather like beaks. All of them in this group are blood-spattered and grubby. They have been tearing at the carcass of a large animal, and all that remains of it is a frightful mess. Lots of them are fighting over what is left, pulling at the meat. They have obviously been feeding for a long time, and most have eaten enough. Many have retired to the bushes nearby but they do not like to see anyone else eating even when they have had enough, so they struggle up and rush at the ones which are still tugging on the sinewy remains. They chase them briefly away and poke at the messy breakfast—only to find that they cannot eat any more. Then they return to the shade to digest.

These savage little dinosaurs hunt in packs. They run at great speed, and the whole flock attacks a victim far larger than themselves, ripping its soft belly open and bringing it down by sheer weight of numbers. They practically eat it alive the way packs of wild dogs do in Africa today. These horrid little meat-eating dinosaurs hiss and splutter at each other while they squabble over the feast, sounding like angry goannas. Let's move on to something less revolting . . .

The hilly country is forested with conifers. Podocarps and kauri pines grow straight and tall, like they do in New Zealand forests today.

We will now change direction and head east, moving away from the inland Eromanga Sea, towards the Pacific Ocean.

As we swing onto our new course a large and active volcano looms up ahead. It is belching smoke and fire, and throwing up showers of rocks, and a river of lava makes an orange channel down one side of its steep cone. In fact, a band of volcanoes can be seen running north-south, some with thin plumes of smoke rising from them, another in the far distance erupting with great ash clouds.

This line of volcanoes would fit between Rockhampton and Townsville on the modern map. In this period in the history of our continent the coastline is much further away to the east, a considerable distance from the range of volcanoes.

The volcano landscape is awful and forbidding. We advance in time by a few hours to see the scene by night in all its fiery majesty . . . the red and orange illumination of the night sky, the black silhouette of the volcanic cone, the glowing orange of the river of lava, and the fireworks display. It all looks like a giant Roman candle, and there is lightning flashing in the clouds above. It is an awesome sight, and a dramatic episode in the formation of the eastern part of Australia.

Back in the daylight and the park-like landscape of Muttaburrasaurus country, is an absolutely enormous

A palm-like cycad

dinosaur. Over there, beside a large billabong. It is every bit of 16 metres long, and at least three metres high at the hips. It must be a relative of *Rhoetosaurus*, and one of the last big Sauropods.

By Early Cretaceous times, big plant-eating dinosaurs of this sort had become extinct in most parts of the world, but they still survived on Australian territory. They must have consumed vast quantities of plant matter. There were no grasses or broad-leaf flowering plants for them to feed on like elephants do today. Instead they ate the large, colourful cones of cycads, the young, unfurling fronds of tree-ferns, and dry, leathery cycad fronds as well as the resinous foliage of conifers.

But this huge creature has been knocking over the tall palm-like cycads and ripping open their trunks to consume the sago starch inside. This is good, nourishing food, and it has the added advantage of swelling up in the stomach like some diet foods do, making the dinosaur feel well fed on quite a small meal. We are going to find out something else about what a Sauropod eats for lunch because this one is entering the water and starting to feed. It snaps its beaky snout and pumps water through its mouth, sieving out the dense mats of algae which float in the water like green porridge.

If you have ever watched a domestic duck feeding you would see the similarity. There was a lot of algae in the shallow waters of lakes and estuaries in Cretaceous times. The oils in the rich oil-shale deposits which exist today were originally formed from these algae. In the modern world they are being mined for processing into petrol. Each minute, unicellular plant had a droplet of oil in it and countless billions combined to form scummy rafts in the water. They proved a source of highly nutritious food for water-loving dinosaurs. As a major part of the Sauropod's diet they resulted in the formation of blubber, and Sauropods are the original big-fat-slobs, all fatty tissue and very little muscle. Add to this the smallness of their brains compared with their vast size, and it seems little wonder that they became extinct.

Further on, we pass another river with deep pools, quiet reaches and small waterfalls—and another burra of Muttaburrasauruses. They are busy acting like hippos and are hard to see in the water. But where the river runs through a gorge nearby, a lone female walks purposefully through the parkland, heading for a rocky scarp. Her body is heavy and distended because she is pregnant. At the moment she is on her way to a safe place for laying eggs—at the foot of a small cliff where she will find the things which she needs for this big event. Water seeping down the cliff has weathered the rock and made a clay patch—orange-brown, sticky clay like we use to make pottery. The mother Muttaburrasaurus starts by making a shallow hole in the clay, scraping with her hands and kicking with her back feet as well. Then she sits in the hollow, folds her tail around herself and revolves, pushing the pliable clay into a smooth, dish-like nest. When it is to her liking, she shoves and pushes earth up round the edges, raising the rim. Who would have thought that a dinosaur would make a neat, round clay nest?

She wanders off along the base of the cliff and pulls moss off in shaggy sheets, carries them in her mouth, and pushes them into the damp clay of the nest, making a soft, springy lining.

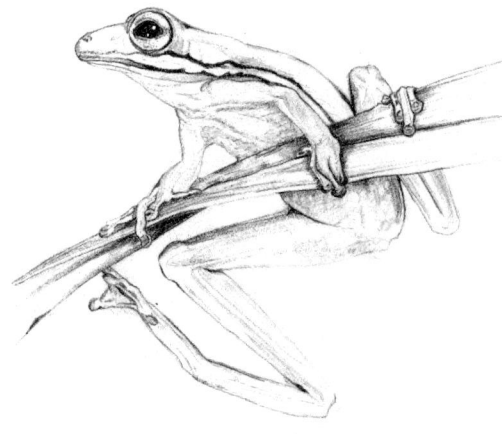

It looks as though she is going to be busy all day, getting the nest the way she wants it, so we walk along the scarp and find that this is, indeed, a maternity ward. There are three other nests in similar situations, each of them containing eggs covered over with moss and bits of fern. The mothers have kicked leaf-litter over them, concealing the eggs as far as possible. The mothers are nearby in the shade, keeping watch. The rocky cliff protects the nests from behind, and the steep slope would make it hard for predators to steal the eggs unnoticed. A mother Muttaburrasaurus intent on defending her young would be a

formidable opponent. Even such a gentle and amiable creature as she usually is can be a fierce attacker if her eggs are threatened.

Something is scratching and scrabbling in the bushes further down the slope. It is not worrying the mother Muttaburrasauruses, so it cannot be anything that endangers their eggs. Whatever it is, is small and hidden by the underbrush, like the butcher shop of little carnivorous dinosaurs was before. But the sounds we hear are different—sort of clucking and peeping noises, rather like contented hens and chickens. And now we see them—a flock of bird-like little dinosaurs, no bigger than frilled lizards, and very much like them, minus the frill. Instead they have a sort of dewlap under their chins, and they are like thin, plucked birds with long, thin tails. They are walking on their hind legs, and scratching and kicking the leaf-litter with their bird-like feet, picking up (or pecking up) the centipedes, beetles, worms and spiders which they disturb. The young ones run behind their parents, peeping and asking to be fed, and flapping their little forelegs which hang like the beginnings of wings. Occasionally the babies are fed insects to keep them quiet.

Seeing this flock it is easy to understand that birds did, in fact, evolve from reptiles. We know that there were already birds in Cretaceous times. Fossil feathers have been found in rocks of the Early Cretaceous age in Victoria, indicating their presence in Australia.

Returning to the Muttaburrasaurus nest, we find that the eggs have already been laid. They are football size and oval, a pale duck-egg blue with a sprinkling of brown spots, and there are seven of them, very neatly arranged in the middle of the nest, all standing, touching each other, pointed end up. They look like Easter eggs in a basket arranged for a children's party. The mother is collecting moss to cover them up and will conceal the nest like the others we have seen when she is sure that the eggs are secure. To find out what happens to the eggs, we have to travel forward in time to fit the whole story together.

The Muttaburrasaurus is leaning over her nest and she is listening intently. We listen too, and we can hear a scrabbly-scratchy sound from inside the nest and a faint didgeridoo chorus. The babies are calling inside their egg shells and are ready to hatch. The mother's excitement increases and she pushes her nose into the nest, blowing the sand and moss away in little puffs, revealing the blue tips of the eggs. In the same way, modern crocodiles watch their nests and hear the young calling before they hatch, and are there to welcome and defend them in their first few hours.

The egg shells split as the babies push and struggle inside, and the mother is now making answering sounds. One by one the eggs crack open and faces appear in the

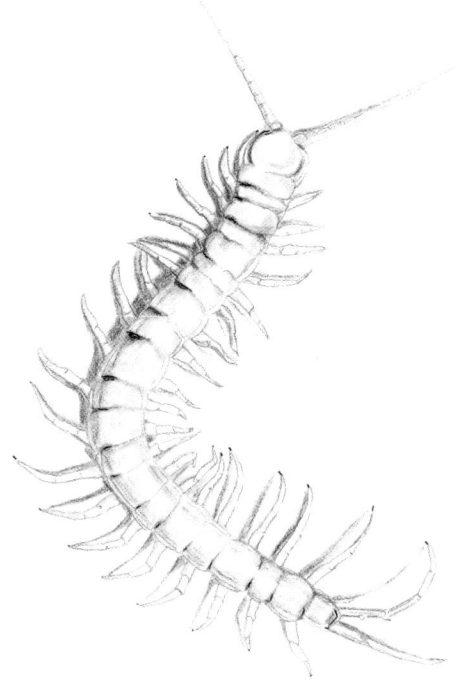

jagged holes as new little dinosaurs look out into the world. They seem to be in no hurry to emerge completely. Maybe they are catching their breath after the exertion of breaking the shells. Then they all push together, and the egg shells fall apart, and they are born. They are coloured like exotic caterpillars, and look very similar, except for their legs and their size. The mother is nuzzling them, and picking away the shell pieces, which she crunches up and swallows.

The seven new dinosaurs settle in the nest and the mother goes off to find them food, with many a backward glance, obviously worried about leaving them. She does not have to go far because the nest site has been chosen with the food supply in mind. The cycads growing in dense thickets here are a special *Pentoxylon* variety and this is their season for fruiting.

They all have trusses of fleshy, mulberry-like fruits, red and ripe, very tasty and nutritious. She picks them off with her beak and stuffs her cheeks full until she can pick no more. Then she munches them up and returns to the nest. The youngsters thrust their snouts into her mouth and gobble like baby birds, fighting each other for a turn.

For the first few days of their lives the babies will stay in the nest and the mother will feed them. As well as fruits they eat cicadas which swarm in the bush at this time of year and make an ear-splitting racket during the heat of the day. They will progress to eating cones of cycads and other vegetable matter which they have to chew themselves, and occasional meat if the mother kills a small bird-like dinosaur. Though Muttaburrasauruses mainly eat plant material, they supplement their diet occasionally with fish, insects, eggs, and even some carrion if they come upon the left-overs of a meat-eating dinosaur's feast.

The young Muttaburrasauruses grow at a great rate and are very soon overflowing the nest. The day comes when the mother decides it is time to return to her burra, taking her family with her. She sits down and the young ones

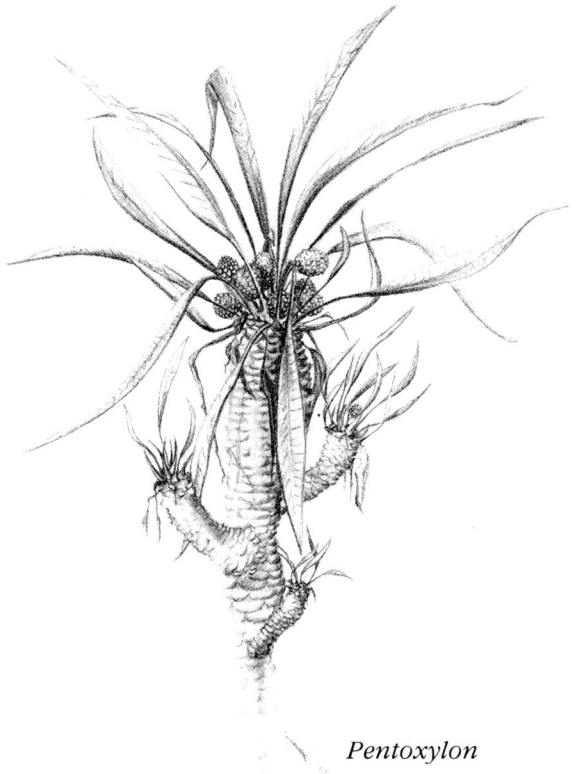

Pentoxylon

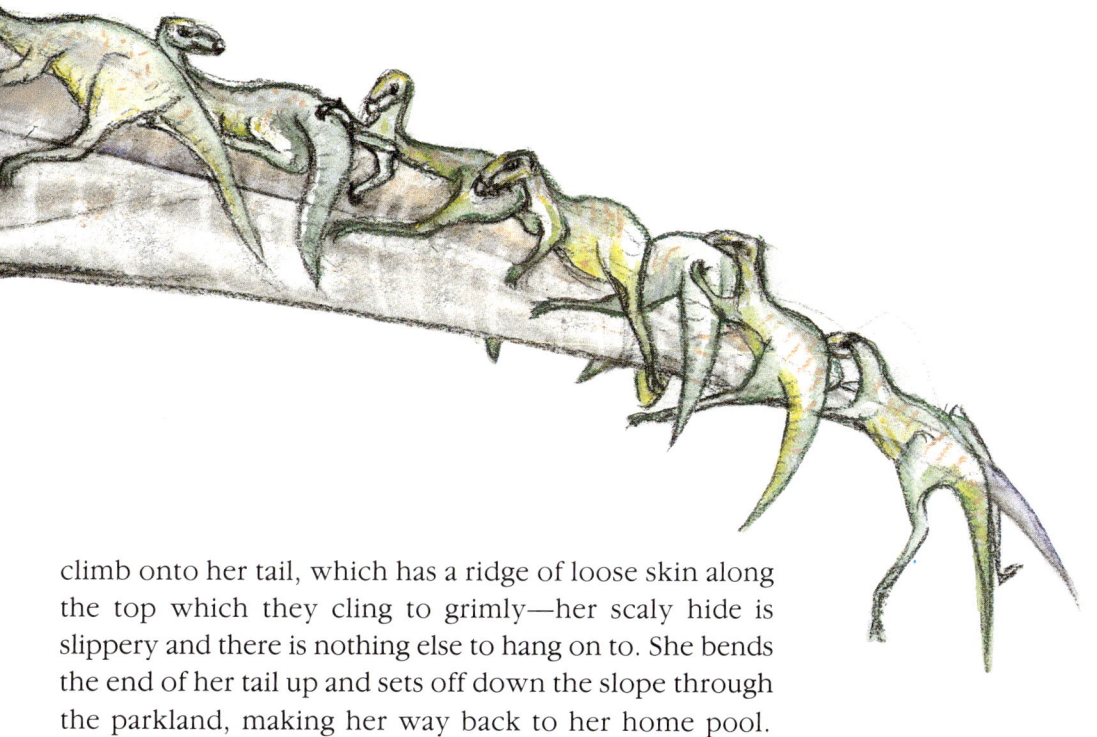

climb onto her tail, which has a ridge of loose skin along the top which they cling to grimly—her scaly hide is slippery and there is nothing else to hang on to. She bends the end of her tail up and sets off down the slope through the parkland, making her way back to her home pool.

The nesting site is a fair way from the burra. Breeding females from many burras all come to this central place to nest. Each burra has a big male and his harem of three or four females, as well as the juveniles of the last breeding female. As soon as the young are half-grown they leave their mother's burra and join communities of many uncles and some aunts, and all the immature males and females. They have much less well defined homes and roam the countryside. Many in these herds will never breed, and

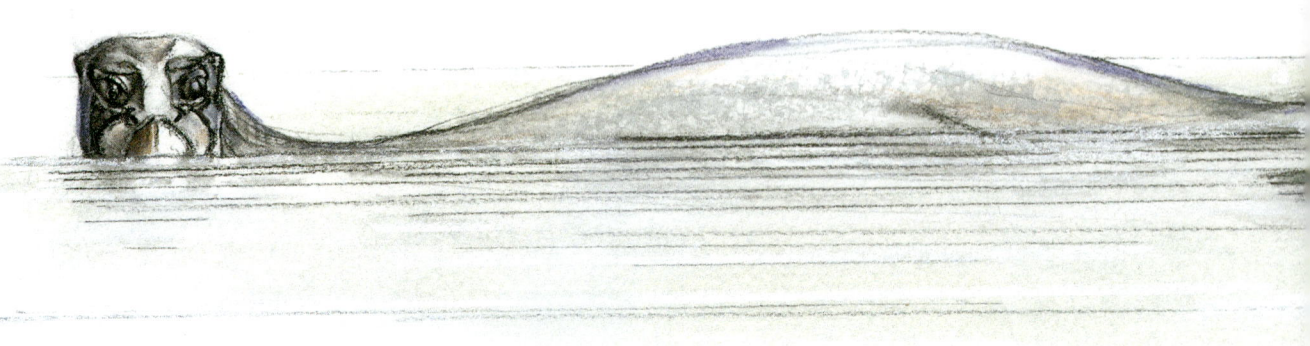

many will fall victim to predators. The dreaded *Allosaurus* still rampages about killing and eating, and he has an immense appetite, and there are many lesser hunters always on the lookout for the young and the unwary. The non-breeding adults look different from the burra members. They are khaki-green and not like the splendid creatures we have seen in their breeding colours. They are better camouflaged for spending their time in the bush, and in shallow pools and swamps, for the most desirable deep pools are occupied by the breeding males and their chosen females.

Our mother Muttaburrasaurus has arrived at her home pool. She slides happily into the deep water, grateful to be able to play hippo again like the others in her burra. Her children float off and, startled by their first introduction to the water, they struggle back and grip their mother's tail flange with the claws of at least one foot. They find themselves floating at the right level—eyes and nose exposed, just like their elders.

Let's now leave these all-Australian dinosaurs and travel further east to the Pacific Ocean.

Here the sea looks just the same in Muttaburrasaurus time as it does in ours. Blue and beautiful, white waves tumbling up the sloping beaches, shining sand, a line of shells and seaweed left by the tide. There are sand dunes rolling inland, and weathered cliffs where the waves are crashing against the rocks and the spray flies high. The thunder of breakers fills our ears, the fresh sea wind cools our faces, and there is a salty smell . . . There is nothing "prehistoric" about this scene.

But wait . . . Those are not familiar birds flying over the water, circling and wheeling, riding the air currents and sliding down the roller-coaster runways of the skies. They are as big as wandering albatrosses, they dive like gannets, dropping like stones, splashing into the sea and rising up again with fish in their beaks. They fly off to the cliffs carrying their catch.

These are *Pterosaurs*—graceful winged dinosaurs. Their long-beaked snouts have sharp teeth, and they look like a cross between huge flying-foxes and birds. They are featherless and furless—a shiny silver-grey below and darker on their backs. There is a rookery on the cliffs, like

a modern sea-bird rookery, where they always nest and roost. The rocks are white with guano (the accumulation of droppings from generations of these creatures). The nests are no more than places on the rock ledges where the pebbles and debris have been pushed aside, leaving space for two white eggs the size of cricket balls, or the nestlings. There are many baby Pterosaurs sitting waiting for their parents to bring them fish. They are very awkward looking. One can't help thinking that they would be far more comfortable if they knew how to hang upside down like bats do. With their wings folded up they are more or less sitting on their elbows at the front. Their front feet are reduced to claws half way along their wing edges.

Very young Pterosaurs feed from their parents' beaks the way young birds do, putting their snouts right into the adult's throat while it regurgitates part of its last meal. They flap and jiggle while they feed, making a mewling, squeaking noise, partly choked by food. When they are older they are brought whole fish to eat, and they throw back their heads and swallow them whole, shivering them down their throats.

In the ocean beyond live enormous and terrible predatory reptiles. One of them, *Kronosaurus*, may be as long as 14 metres, and there are others as big as the Loch Ness Monster, though not as harmless. But none of these can be seen from the cliffs—the age-old sea looks just like our sea does today.

Now it is time to journey back to our own time and space from the day of the Muttaburrasaurus long, long ago . . . long, long ago.

FACTS ABOUT THE MUTTABURRASAURUS
Muttaburrasaurus langdoni

A skull and most of a skeleton were discovered in Early Cretaceous marine deposits on the Thompson River, near Muttaburra in Queensland, about 100 kilometres northeast of Longreach in 1963. Another skull was found in August, 1987.

CLASS	*Reptilia*
ORDER	*Ornithischia* (meaning bird-hipped)
SUB-ORDER	*Ornithopoda* (meaning bird-footed)
FAMILY	*Iguanodontidae*

MUTTABURRASAURUSES were able to walk on their back legs, but probably spent much of their time browsing on all fours. Studies of their teeth suggest that they were largely herbivores (plant eaters), but were probably also partly carnivorous (meat eating).

DINOSAURS are known to have made nests, and fossil eggs have been preserved, a few even with baby dinosaurs inside.

About 500 kilometres west of Muttaburra, near Boulia, Queensland, Pterosaur fossils have been found in Early Cretaceous sediments.

AUSTRALIA, IN EARLY CRETACEOUS TIMES was partly flooded by sea. The areas which were land are coloured green, and the sea is blue. The modern outline of the continent is also shown.

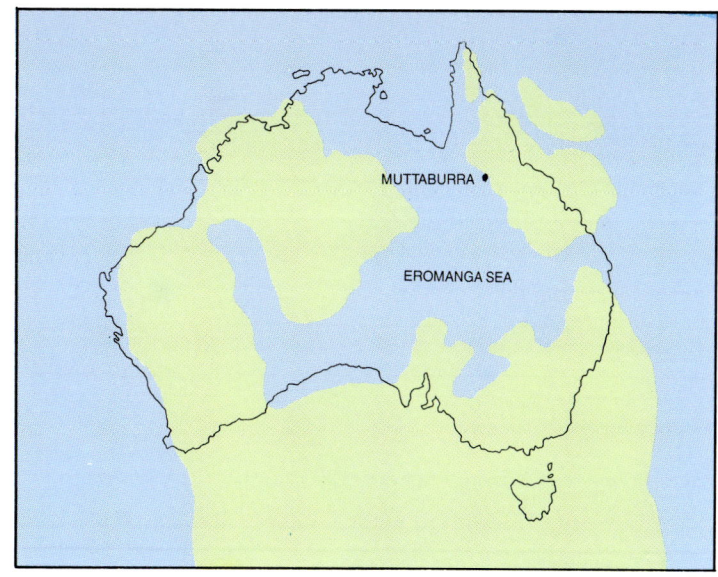

GONDWANA, THE GREAT SOUTHERN SUPERCONTINENT broke up into the modern continents (coloured pale green on this map).

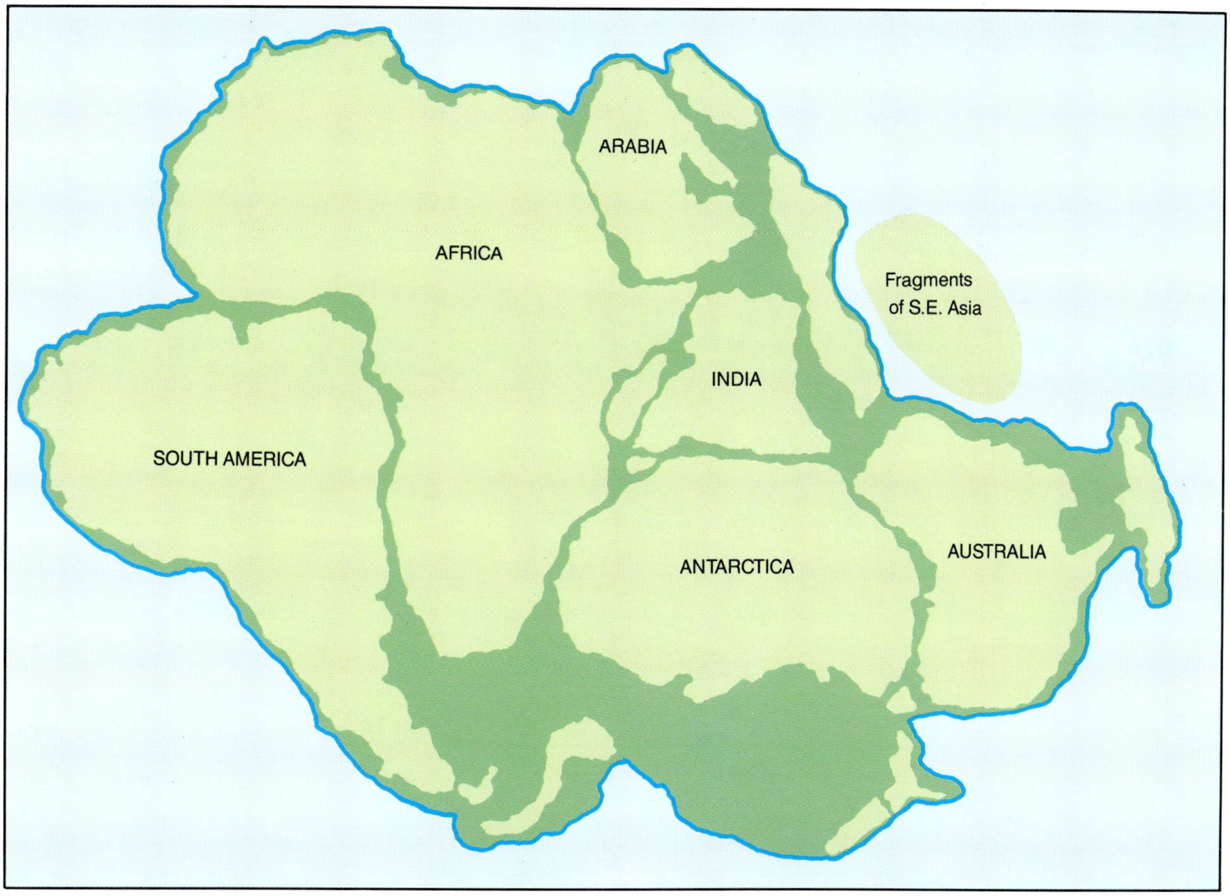

First published in 1990 by
Houghton Mifflin Australia Pty Ltd
PO Box 289, Ferntree Gully, Victoria 3156
112 Lewis Rd, Knoxfield 3180

Produced by
John Ferguson Ltd.
59 Elizabeth Bay Rd
Elizabeth Bay, NSW, 2011

Copyright © Mary White
Illustrations © Robyn Muche

All rights reserved. No part of this publication may be
reproduced, stored in a retrieval system, or transmitted in any
form or by any means, electronic, mechanical, photocopying,
recording or otherwise, without prior written permission of
the publisher.

National Library of Australia
Cataloguing-in-Publication Data

White, M. E. (Mary E.).
  Muttaburrasaurus

ISBN 0 86770 100 5.

  1. Dinosaurs — Australia — Juvenile literature.
  1. Title. (Series: Prehistoric natural history series).

567.91

Designed by Jane Tenney
Typeset by Midland Typesetters, Maryborough, Victoria
Printed in Hong Kong by South Sea International Press Ltd.

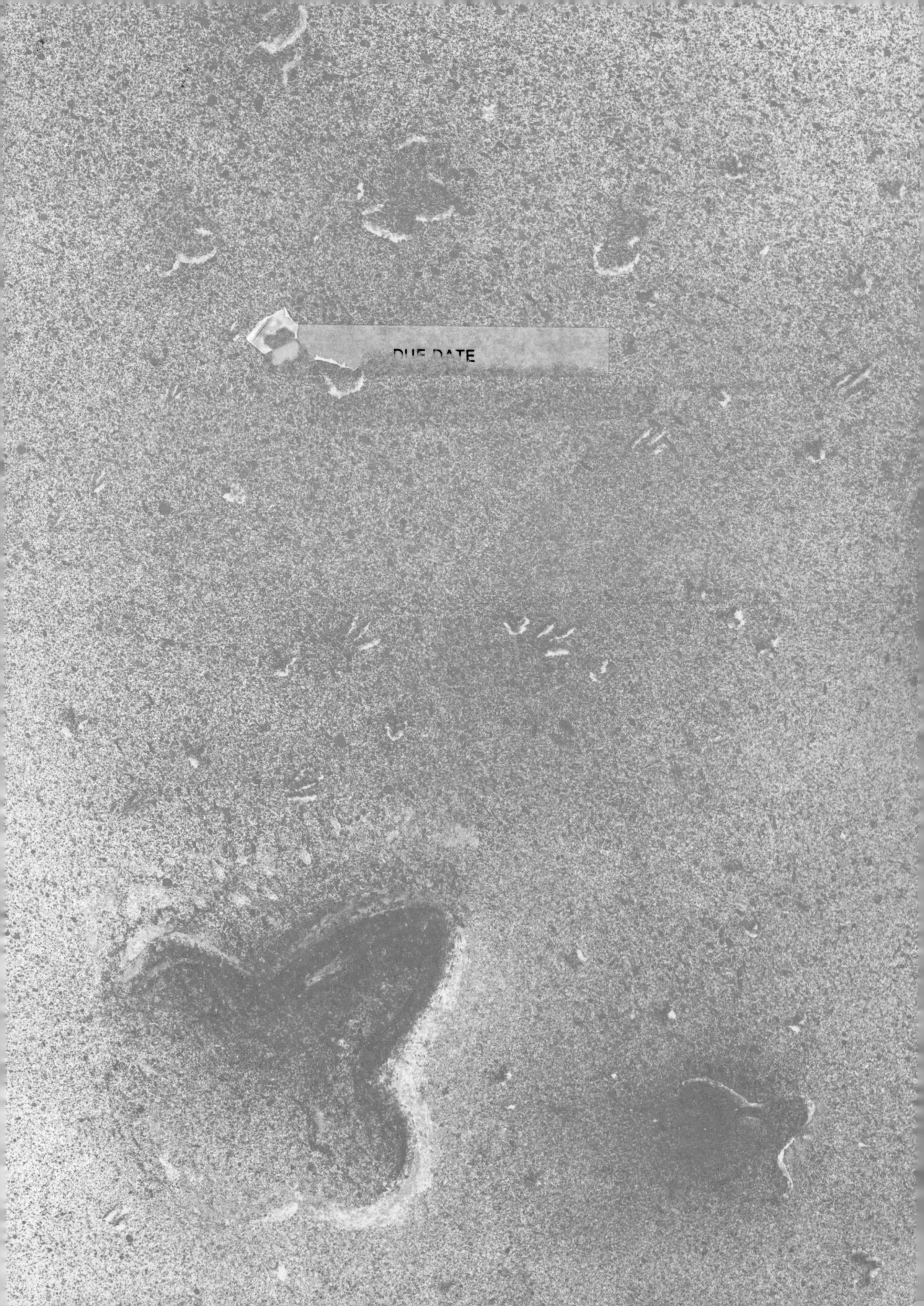